# A ROCKY MOUNTAIN
# LICHEN PRIMER

# A ROCKY MOUNTAIN
# LICHEN PRIMER

James N. Corbridge
*Professor of Law*
*University of Colorado School of Law*

William A. Weber
*Professor Emeritus*
*University of Colorado Museum*

Photographs by
*Ken C. Abbott*
*University of Colorado at Boulder Public Relations*

UNIVERSITY PRESS OF COLORADO

Copyright © 1998 by the University Press of Colorado
International Standard Book Number 0-87081-490-7

Published by the University Press of Colorado
P.O. Box 849
Niwot, Colorado 80544

All rights reserved.
Printed in Hong Kong.

The University Press of Colorado is a cooperative publishing enterprise supported, in part, by Adams State College, Colorado State University, Fort Lewis College, Mesa State College, Metropolitan State College of Denver, University of Colorado, University of Northern Colorado, University of Southern Colorado, and Western State College of Colorado.

The paper used in this publication meets the minimum requirements of the American National Standard for Information Sciences — Permanence of Paper for Printed Library Materials. ANSI Z39.48-1984

*Library of Congress Cataloging-in-Publication Data*

Corbridge, James N.
    Rocky Mountain lichen primer / James N. Corbridge, William A. Weber : photographs by Ken C. Abbott.
    p. cm.
    Includes bibliographical references (p.  ) and index.
    ISBN 0-87081-490-7 (pbk. : alk. paper)
    1. Lichens—Rocky Mountains Region—Identification. 2. Lichens—Rocky Mountains Region—Pictorial works. I. Weber, William A. (William Alfred), 1918– . II. Title.
QK587.5.R63C67    1998
579.7'0978—dc21                                97-48870
                                                        CIP

07 06 05 04 03 02 01 00 99 98    10 9 8 7 6 5 4 3 2 1

Lichen displayed on cover is Caloplaca trachyphylla, Dinosaur National Monument, Utah.

It is by studying little things that we attain the great knowledge of having as little misery and as much happiness as possible.

—Samuel Johnson, in Boswell's *London Journal, 1762–1763*

The defeat of project Chariot [a plan to create a huge harbor in northern Alaska with the aid of up to six thermonuclear bombs] represented the first successful opposition to the American nuclear establishment and one of the first battles of the new era of "environmentalism." Here the rationale for caution was not the old logic of conserving a magnificent landscape or endangered species. Rather it was based on a more holistic concept of environmental protection, which recognized that insidious degradation was possible because of the invisible connectedness of things. "**Looking back on my career in environmentalism,**" said Barry Commoner in a 1988 interview, "**it is absolutely certain that it began when I went to the library to look up** *lichen* **in connection with the Chariot program. That's a very vivid picture in my mind.**" Chariot led Commoner into environmentalism, and Commoner led others into what became known as the environmental movement. "**And I think,**" said Commoner, "**in so far as I had an effect on the development of the whole movement (which I did, I have to admit), Project Chariot can be regarded as the ancestral birthplace of at least a large segment of the environmental movement.**"

—Dan O'Neill, *The Firecracker Boys*, 1994

# CONTENTS

| | |
|---|---|
| Preface | *ix* |
| Introduction | *xi* |
| Illustrations and Text | *1* |
| Additional Sources | *39* |
| Glossary | *41* |
| Index | *45* |

# PREFACE

The idea for this little book came about as follows. On Saturday mornings over several years I and a few of my colleagues gathered in the herbarium for coffee. One of us was Jim Corbridge, then the Chancellor of the University of Colorado at Boulder, and Professor of Law. I had inherited a large number of classical wooden jigsaw puzzles (Pastimes) from my aunt, most of them almost a hundred years old, and there was always one on the table waiting to be assembled. Jim and Ron Wittmann, a physicist friend at the National Institute of Standards and Technology, amateur botanist and now co-author of my revised *Colorado Flora*, were addicted to the puzzles, so much so that the work to which Ron and I dedicated our weekends began to suffer.

One day, as I gradually weaned my friends away from their favorite pastime, Jim asked: "What is all this about lichens?" I began to explain to him some rudiments about lichenology, and before I knew it Jim was converted, delving into all the literature he could get hold of. Soon we were looking at lichens together, going into the field, and Jim began to make his own little collection of lichens. It was so much fun for him that he proposed that we put together a little identification guide, not one to produce professional lichenologists, but one that would introduce the common lichens to rock gardeners, hikers and climbers, requiring a minimum of technical terminology, and showing how the most common and conspicuous lichens can be recognized on sight without recourse to the microscope. The result is this lichen primer.

The color photographs are taken from specimens housed in the University of Colorado Herbarium. We are greatly indebted to Ken Abbott, photographer with the University Public Relations Department, for his excellent capture of the essence of the species and to Ron Wittmann for help with layout and typography.

Looking at some of the photographs of crust lichens, we suspect that we have substituted living jigsaw puzzles for the old ones! We hope that you will enjoy their challenges.

—W. A.W., December, 1997

# INTRODUCTION

Lichens are an important, if overlooked, part of the plant kingdom. Over 15,000 species have been identified worldwide, and they grow in all parts of the earth, from the tropical jungles to the extremely hostile environments of Antarctica, the alpine tundra, and bare lava flows. They are attached to a variety of substrates, including the soil, bark, mosses, and rocks — old gravestones are a common site — and cliffs and desert landscapes can be drenched in the chartreuse, yellow, green, or orange of various lichen species. A few blow about unattached to anything.

Throughout history, lichens have served as sources of dyes, for medicinal purposes, and as forage. They are a staple of the caribou diet. Because of some lichens' extremely slow growth rate (some living thalli are up to 10,000 years old), scientists are using them to date glacial recessions. Many lichens are sensitive to air pollution and are being used to determine regional air quality.

Lichens are a conspicuous part of the scenery of the Rocky Mountains, particularly those occurring on rocks. It is natural that outdoor people see them and want to know about them as organisms, and to find out what their names are. This book presupposes no prior knowledge of lichens, and aims at recognition of the most commonly encountered kinds, using only the eyes and the help of a hand lens, which can be acquired at most camera stores. Serious study of lichens can be very time-consuming and frustrating, requiring the help of an expert, access to an excellent collection (herbarium) of specimens, and extensive literature in many languages. But there is no reason for laymen to bother themselves with studies in depth; many lichen species are very easy to recognize and not very variable; anyone can master the recognition of those dealt with in this book.

## WHAT IS A LICHEN?

A lichen is an intimate association of a fungus and an alga. The fungus is the part that we see, while the algae are hidden just under the surface. If you break a lichen thallus, you can see with a lens the thin green layer of algae. What is this intimate association, often called symbiosis? Living within the fungus body, algae can grow in places otherwise impossible for them, and the fungus, by using the algal cells for nourishment, can do likewise. Algal cells, of course, are killed in the process, but they reproduce as one-celled organisms and reproduce faster than they are destroyed. We can call this a very efficient parasitism in which the parasite does not destroy the host.

Despite the many lichen species present in nature, their development remains a mystery. How can wefts of fungus fluff (hyphae) form all the shapes they do and remain different and unchanging—sheets, little bushes, hanging drapery, cups, or shields? They have no known "organizers" as do the growing tips of higher plants, yet they form dichotomous branches, upper and lower surfaces, special reproductive structures, and special hairs that fasten them to the substrate. They can dissolve the cement that holds rock particles together, and they range in life expectancy from a few years to centuries. Lichens have been separated in the laboratory into their components and grown in test tubes, but their restitution as lichens is extremely difficult because life as lichens depends on a starvation diet, and efforts to feed either partner upsets the delicate balance and the fungus becomes a killer of the alga. Growing lichens as lichens is a science in its infancy.

While many types of fungi have been found associated with algae, most come from the class called ascomycetes (cup fungi, which include yeasts); a relative few are basidiomycetes (which include mushrooms). The algae are either green algae (many genera) or blue-green algae (now called cyanobacteria). Lichens containing green algae do not get their colors from the algae but from pigments; they may show their algal green color when wet. Lichens containing blue-green algae often are gelatinous when wet and crisp when dry, black or dark green.

Unless you are willing, literally, to dig into a lichen, it will remain mysterious. Actually, the structure of a lichen is simple. And, although lichens consist of wefts of fungal threads (hyphae) intermixed with some algal cells, they are extremely similar to another plant form which uses different materials, but which we all know something about — a green leaf!

The body of a lichen is called the **thallus**. Take a sharp pen-knife or razor blade and make a few vertical slanting cuts across a lichen thallus.

You will see, below the colored upper surface, even with the naked eye, a thin line of bright green. This is the **algal** (now called the **photobiont**) layer. Although a lichen is two different organisms growing together, the form of a thallus is remarkably similar to that of a green leaf. A leaf has a colorless upper layer (**epidermis**). The lichen has a usually colored upper layer (**cortex**); it consists of densely packed fungal hyphae. A leaf has a layer below the epidermis, of vertically-elongated green cells closely packed together (the **palisade layer**). The lichen has green algal or bluegreen algal (cyanobacteria) cells. The fungus feeds on these green cells, but they reproduce faster than they are consumed. The leaf has a lower layer of pale cells forming a very loose network (the **spongy mesophyll**). The lichen has a loose organization of fungal hyphae (**medulla**). There may or may not be a lower epidermis (in the leaf) or a lower cortex (in the lichen). Leaves have stomates (doughnut-shaped pairs of cells that open and close, allowing the exchange of oxygen and carbon dioxide). Lichens do not require openings, but frequently have cracks or holes (**pseudocyphellae**) in the cortex through which hyphae and algae may poke. The lower cortex often produces hairlike **rhizines** or suction-cup stalks (**hapteres**) that anchor the lichen to the substrate.

The arrangement of the various layers in the lichen and the leaf are extremely slimilar, and is the most efficient for the process of photosynthesis. The fact that the end result is obtained from completely different materials is one of the wonders of plant life. For the most part, it is thought that the spores are very infrequently involved in the reproduction of the lichen, but are leftovers from the days long ago when the fungus lived independently. Try to imagine how this curious relationship between fungus and alga began, and how the various lichens evolved over millions of years. This is the great question for which no one yet has the answer.

**Apothecia** are very important in the recognition of many lichens. They are usually small, rimmed cups or disks, appearing on the surface of the thallus. The margins are sometimes, but not always, the color of the thallus. The disks themselves are almost always a different color. **Perithecia** also produce spores, but these structures, flask-shaped, with a small opening at the top, are imbedded in the thallus. Apothecia and perithecia produce spores (produced by meiosis) which are diagnostic of species, but they probably play an insignificant role in reproduction, because the lichen must reproduce by fragmentation of a thallus containing both fungus and alga. The strains of algae found in lichens evidently do not occur free-living in nature.

A few lichens produce special spores in surface structures known as **pycnidia**. These are buried in the thallus, appearing as tiny black dots on the surface where the spores escape. The pycnidial spores are formed by budding. The shapes and sizes of these spores usually are characteristic of a genus.

Lichens usually reproduce by distributing clumps containing algal cells and fungal hyphae. The presence of each is necessary for this type of reproduction. Such clumps include **isidia**, small growths with cortex and medulla that can break off and propagate the lichen, and **soredia**, powdery clusters, lacking cortex, that break through the surface of the thallus.

This book is concerned with recognition of some of the common and conspicuous lichens in the field. Although the examples are taken from the Rocky Mountain region, most of these lichens occur in mountain areas throughout the world. For convenience, lichens can be divided into four categories: **crustose, squamulose, foliose,** and **fruticose**.

**Crustose** lichens are closely anchored (appressed) to the substrate. They may appear almost "painted" on the substrate, and because of their close attachment they are difficult to collect unless part of the substrate is taken as well. The thallus of a crustose lichen has no lower cortex, and consists of an upper cortex, an algal layer, and a medulla which functions to attach the lichen to the substrate. Some crustose species are areolate, cracked into small islands or areoles. See, for instance, *Rhizocarpon geographicum* (Plate 3).

**Squamulose** lichens have a thallus made up of scalelike units or lobes (squamules), for the most part closely attached, but free or upturned at the edges. A good example is *Psora decipiens* (Plate 12). A strikingly different squamulose lichen is the genus *Cladonia*, which often appears on soil, moss, or rotting wood such as old fence posts or fallen trees. Its thallus sends up fruiting stalks, called **podetia**. In some species these may be branched, or may look like tiny wine goblets or golf tees. In others, the stalks may be straight posts, perhaps a quarter of an inch tall. For examples, see *Cladonia ecmocyna* and *C. pyxidata* (Plates 27 and 28).

**Foliose** lichens are leaflike or have distinct, often large, lobes. They are attached more loosely at the substrate. The thallus of most foliose lichens has both an upper and lower cortex, and many have hairlike structures called **rhizines**, which grow from the lower cortex and serve to attach the lichen to the substrate. Some foliose species are attached to a central umbilical point from which the lobes radiate. For illustrations, see *Parmelia, Umbilicaria,* and *Dermatocarpon*.

**Fruticose** lichens tend to be more upright and only basally attached to the substrate, and have a wide variety of forms. They have no upper and lower surfaces and are generally cylindrical in shape, like a tree or shrub. The commonly encountered genera, *Bryoria* and *Usnea*, have a branched thallus and look like a miniature bush or beard. Some of these hang from branches, resembling Spanish "moss".

## A WORD ABOUT COLLECTING

Once you learn to identify the more familiar lichens in the field, you may want to start a modest collection of your own. Begin by acquiring a few inexpensive pieces of equipment to accompany your hand lens. A rock hammer and cold chisel, available at most hardware stores, will enable you to chip away pieces of rock, leaving the lichen intact. Rocks shatter and hammers sometimes miss the target! A gardener's glove for the hand holding the chisel, and safety goggles, are essential. Strive for as thin a piece of rock as you can chisel off without destroying the attached lichen.

Next you will need a sturdy straight knife with sheath. Keep it sharp! A dull knife is more dangerous to use than a sharp one. Don't have a folding knife; it will fold on your finger while you are working. The knife will assist you in removing foliose and fruticose lichens from the substrate, or in collecting lichen-bearing bark with the lichen still attached. Sometimes the lichens will be dry and crumbly. A bit of moisture will make them much easier to collect. Use a small spray bottle, available at the local hardware store and filled with tap water. Put this equipment in a cloth grocery bag, add your hand lens, and you will be ready to go.

When you have collected a specimen, keep a record of the date and place you found it. Also note the material it was attached to, as this may be helpful later for accurate identification. Put each specimen in a paper bag. Plastic sacks are not so good, as they retain moisture which may cause the lichen to mold. When you get home, you can transfer your lichens to individual paper envelopes for future reference. They will last a long time. Some of the lichens collected by Linnaeus, the famous 18th century Swedish botanist, are still identifiable today.

## THE LATIN NAMES

One of the imagined difficulties of learning to know the lichens comes from the use of Latin names. This should not be a problem! It is true that in the United States, unlike some other countries, no systematic effort has been made to assign common names to the lichens. Such common names as there are may be complicated or ridiculous, and the scientific names are simple and often meaningful. A modest amount of Latin need not be intimidating. Such Latin words as stadium, militia, and alma mater are mastered without suspicion; children readily rattle off the scientific names of the dinosaurs to which, with rare exceptions, common names have not been given. Armed with the scientific names, you will identify the lichens you know wherever you encounter them in your travels. Consistent with uniform botanical practice, the first, or generic name indicates the genus, while the second word (specific epithet), coupled with the generic name, makes up the specific name. In this book, the English meanings of specific epithets are given when they are known.

## THE CARE AND FEEDING OF LICHENS

With the population movement into the Rocky Mountain region, lichens have become desirable adjuncts to the exterior walls and fireplaces of new homes, and especially to rock gardens. These so-called "moss rocks" are taken from boulders and layers of rock formations and sold at high prices, exorbitant when one realizes that the lichens on them are not indestructible. The movement of lichen rocks out of their natural places should be thought of as a conservation problem, for once removed it takes centuries for them to replace themselves. Gardeners and home owners should be informed that these lichens can be killed by care. They do not require watering, they grow very slowly (few people will be able to live long enough to see any growth), and they are sensitive to pollution of several kinds. In the home, lichens on the fireplace, if watered, may be quickly destroyed by the molds that are prevalent in the air of the rooms. When they accumulate dirt, they are difficult to clean. Some crusts may be cleaned by wadding up white bread and using this as a "wallpaper cleaner." Foliose or fruticose lichens are extremely fragile; perhaps a soft air-brushing might help. Do not attempt to make them grow, despite what you might be told by ignorant dealers. A *New York Times* garden editor once suggested that spreading peanut butter over the lichens would help them to turn green. This change is nothing more than a colony of the green mold, *Aspergillus!* Expensive lichen rock walls, inside a covered mall, have been swabbed down and rubbed with wet mops! They didn't last very long. If you have lichens, do not water them; in a city, the automobile fumes will eventually destroy them, and normal watering of a lawn, hitting the rocks, may result in loss of the lichens, which, after all, are nothing more or less than fungal hyphae, "cotton fluff"!

# A ROCKY MOUNTAIN
# LICHEN PRIMER

# ILLUSTRATIONS AND TEXT

**Note:** The names in parentheses are generic synonyms. Lichen names are being changed often nowadays, mostly as a result of refined techniques of analysis of the apothecia and asci and new understanding of relationships.

1. **Pleopsidium (Acarospora) chlorophanum** [Greek, visibly green]. Thallus marginally lobed; apothecia somewhat paler. Makes great chartreuse-green splashes on vertical cliffs and boulders. Can be seen from miles away. Why does it only grow on vertical surfaces? The name *Acarospora* is derived from *Acarinae*, ticks, because the spores in this genus are produced in great numbers in a single ascus, and are minute, about 3–4 micro-millimeters.

2. **Candelariella rosulans** [making small rosettes]. Most crustose yellow lichens belong to *Candelariella*. This one, common through the southwest, has a distinctly clumped (rosulate) thallus. All visible parts are the same shade of egg yolk yellow. Most grow on rocks, but a few occur on bark and dead wood. This species was first described from Flagstaff Mountain, Boulder, Colorado. Another species (*C. spraguei*) grows on seepage lines on rocks, and has a brighter chrome yellow thallus.

3. **Rhizocarpon geographicum** [map-like]. The Norwegians call this *kartlav* (map lichen). These chartreuse yellow and black lichens are abundant all over the mountainous world, and their species are often very difficult to distinguish from each other. They are extremely slow growing, and are used by geologists to date the recession of glaciers and to ascertain the age and the growth curves of the lichens themselves.

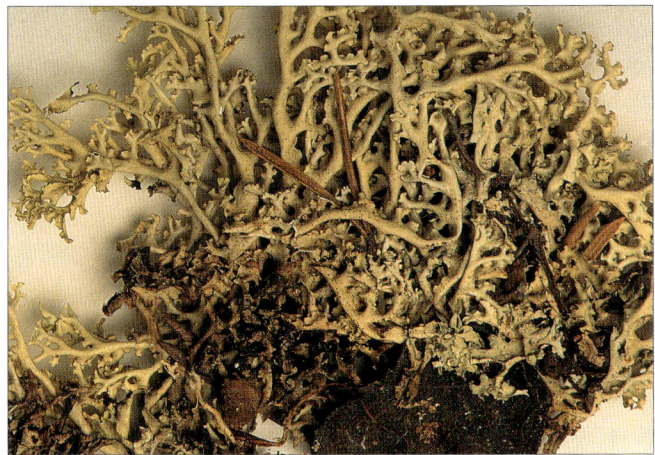

4. **Flavocetraria (Cetraria) cucullata** [hooded]. An erect-foliose lichen growing loose on the soil of the tundra or caught up in masses of dry sedge leaves. The pale yellow thallus is flat, without a distinction between the upper and lower cortex, and the edges are curved inward and not flat as in *A. nivalis*, with which it often grows. A sure field character is the purple base of the thallus. These species never fruit, at least in our region.

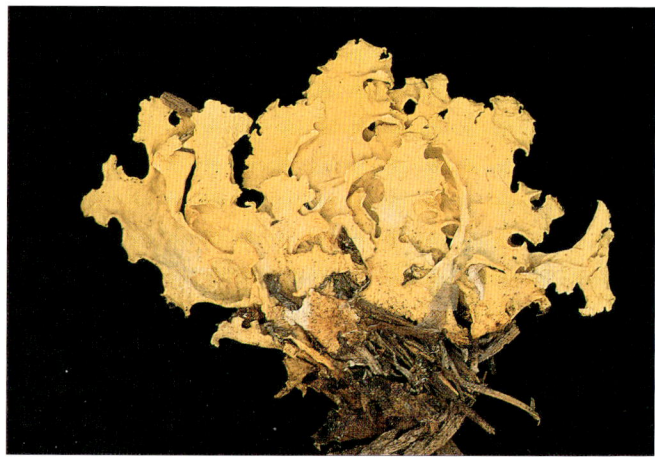

5. **Flavocetraria (Cetraria) nivalis** [of the snow]. An erect-foliose lichen growing loose on the soil of the tundra or caught up in masses of dry sedge leaves. In contrast to *A. cucullata*, the thallus is flat, and the base is yellow rather than purple. These species grow everywhere in the world where there is alpine or Arctic tundra.

6. **Vulpicida (Cetraria) tilesii** [for Wilhelm Gottlieb Tileseus von Tilenau, 1769–1857, Alsatian botanist]. A small bright yellow, erect-foliose lichen growing in tight masses on the alpine tundra. This usually occurs together with *Allocetraria, Thamnolia,* and *Dactylina*. It never bears apothecia, but small black dots on the margins are pycnidia.

7. **Letharia vulpina** [of foxes; this was used to poison animals]. A stiffly fruticose chartreuse green lichen occurring on bark and wood of conifers, particularly cedars. The thallus is angular, and the tips of the branches are black; soredia occur where the cortex is broken. Apothecia are unknown in America and in northern Eurasia where the species spread rapidly after the Ice Ages.

8. **Xanthoria elegans** [elegant]. Very abundant on rocks and cliffs, especially around hunting perches and seepage lines below rodent latrines (tolerant of or requiring nitrogenous compounds). A streak of *Xanthoria elegans* can usually be followed up the rock to a deposit of animal scat. In areas of very high rainfall, the lichen may only be found where large accumulations of animal scat and urine deposits occur, such as the bird rocks off the coast of Scandinavia.

9. **Xanthoria fallax** [deceiving, false]. Abundant on the moist sides of street trees in the towns, and on country roads where horses and cattle stir up nitrogenous dust. One of the most pollution-resistant lichens. Note the underside covered with soredia.

10. **Xanthoria montana** [many-fruited]. A small foliose orange lichen growing on twigs, especially of broadleaf trees, and on the scars of aspen trunks. Apothecia have bright orange discs with a thalline margin, and there are no soredia. The variation from place to place is remarkable, suggesting that several different species are going under this name. The orange pigment in *Caloplaca* and *Xanthoria* is called parietin, and if a drop of potassium hydroxide is dropped on the thallus, it turns deep purple.

11. **Caloplaca trachyphylla** [rough-leaf]. A beautiful rosette-forming orange crust on sandstone, differing from *Xanthoria elegans* in having a coarse, almost warty thallus with tightly appressed, contiguous marginal lobes. *Caloplaca decipiens* grows in similar sites but is more yellow in color and differs by having masses of soredia toward the center of the thallus.

12. **Psora decipiens** [deceitful]. A common desert soil lichen (also occurring on tundra soils) easily recognized by its brick red thallus and black convex apothecia. It is a common component of soil crusts (See also *Psora icterica*). These crusts take many years to form, and are quickly destroyed by trampling, after which the loose soil can be blown away.

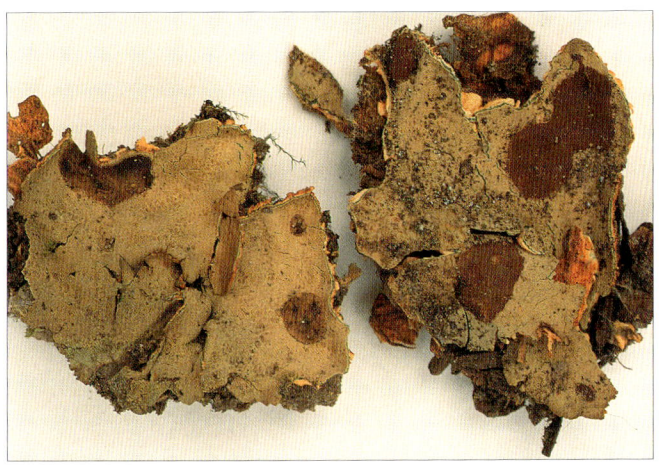

13. **Solorina crocea** [saffron-orange]. An unmistakable brown foliose lichen growing on tundra soil, appressed to the substrate except where the lobes curl up at the edges to reveal a brilliant orange underside. This extraordinary lichen also has two algal layers, a green algal layer just below the cortex, and a layer of blue-green *Nostoc* algae chains just below it. The apothecia are large, flat, brown, and are embedded in the surface level. Three other species occur on the tundra, all with gray thalli.

14. **Psoroma hypnorum** [of mosses]. A brown crust growing on moss or vegetable detritus on the forest floor. The apothecia disk is reddish-brown and the margin is not smooth but with crowded minute gray lobules.

15. **Lecidea atrobrunnea** [black-brown]. A very common species growing on granitic rocks. The brown thallus areoles contrast with the black apothecia, which have a narrow black rim (proper margin or exciple). When well developed, the thallus also has a narrow black border (prothallus), which consists only of fungus tissue, forming a leading edge for future growth. Ranges from the outer foothills up into the alpine tundra.

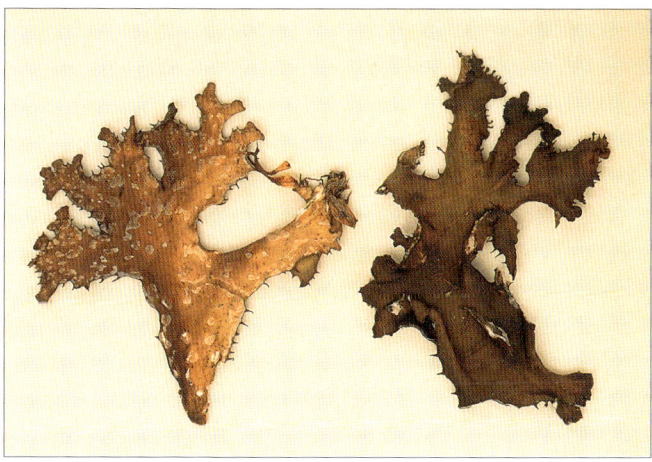

16. **Cetraria islandica** [of Iceland]. An unattached brown, erect foliose lichen with many marginal "spines". The underside is pale tan, and has irregular blotches (pseudocyphellae) where the cortex is lacking. The thallus is very brittle when dry, as are so many foliose lichens, but when wet it can be squeezed tightly, releasing some of its moisture but not being injured. Abundant on alpine tundra but occurring at lower altitudes on north-facing slopes of ravines.

17. **Melanelia (Parmelia) subolivacea** [somewhat olive-colored]. The old genus *Parmelia* has recently been broken up into a large number of genera, of which this is one of the so-called "brown Parmelias". This species is very abundant on the bark and twigs of conifers. A few other species occur on conifer bark, and several are found on rocks.

18. **Nephroma parile** [equal]. A brown lichen with soredia along the margins and on the surface. Resembles *Peltigera*, but the apothecia are borne on the undersides of the thallus lobes, and the underside is bare (lacking veins).

19. **Peltigera rufescens** [reddish]. This *Peltigera* is very common in dry forests. It never seems to produce apothecia, and the thallus is thick and rigid, gray-brown, the lobes with upturned margins. The algal layer is blue-green, hence the thallus does not become green when wet. A gray cobwebby weft occurs on the younger parts of the thallus. The underside has a network of raised veins, with white rhizines occurring along them.

20. **Peltigera venosa** [veiny]. A neat little *Peltigera* containing green algal cells, hence green when wet. The apothecia are circular, on the ends of the broad lobes. The underside of the thallus is white with conspicuous brown veins. Commonly occurring on naked soil banks along trails and forest roads.

21. **Bryoria (Alectoria) chalybeiformis** [steely]. Long, sparsely branched strings of dark brown or black "horsehair". This species grows on rocks, but several others occur hanging from the twigs of conifers. In the Rocky Mountains, wherever *Bryoria* or *Usnea* occur, their presence usually indicates that a late snow patch might be found beneath, giving a moisture evaporation column upward to enhance germination of the lichen.

22. **Coelocaulon (Cetraria) aculeatum** [sharp pointed]. A small, unattached, bushy-branched, dark brown, fruticose lichen with hollow, cylindrical branches ending in sharp points. Apothecia hardly ever seen. Common in gaps in the vegetation, or hidden between other plants, on the alpine tundra.

23. **Dactylina madreporiformis** [name from a genus of tropical corals]. A curious fruticose lichen of the alpine tundra. The thallus is erect, hollow, and with a papery consistency. Apothecia are extremely rare. The lobes on this species are cylindrical. On another species, found in the Arctic (*D. arctica*), the thallus is broader and inflated, like a mitten. Our species is relatively rare in the Arctic and principally a plant of the temperate mountain ranges.

24. **Cladonia cariosa** [decayed]. *Cladonia* is a unique genus, in which the thallus is divided into two distinct parts. At the base, the thallus consists of flattened squamules with an upper and lower side. Arising from these are erect, cylindric stalks (*podetia*) which bear the apothecia at their tips. Apothecia may be red, brown, or rarely tan. This species has squamules on the podetia, and brown apothecia.

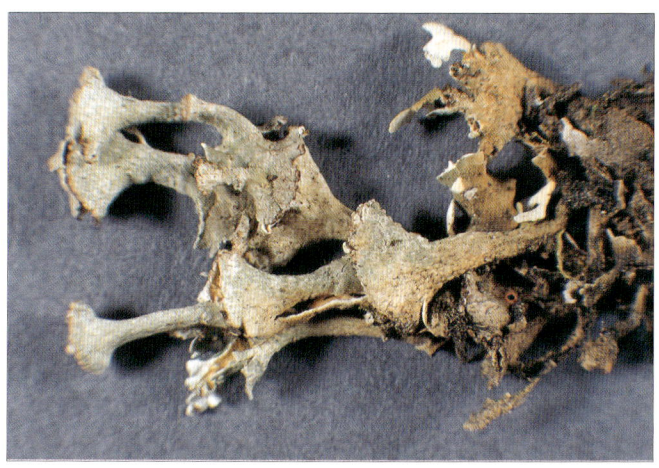

25. **Cladonia cervicornis** var. **verticillata** [antler, whorled]. An unmistakable *Cladonia* in which the podetia are tiered, with flaring saucers arising at intervals along their lengths. A common species of forest floors.

26. **Cladonia deformis** [deformed]. A yellowish *Cladonia* (containing usnic acid) with an irregularly shaped, often slit and fissured, thallus covered with minute soredia. Two almost identical species occur, this one which does not fluoresce in short wave ultraviolet light, and *C. sulphurina*, which does.

27. **Cladonia ecmocyna** [dog-leash??]. Here the podetia are elongate, and have scattered squamules along their lengths. Apothecia are brown, and are borne on narrow cup margins. A close relative, *C. gracilis*, is distinguishable only by chemical tests. Common on forest floors.

28. **Cladonia pyxidata** [with a lid]. Several species of *Cladonia* produce podetia which terminate in deep or shallow cups. This one has well-developed broad cups that are filled with small green squamules, like eggs in a birds nest. The apothecia are brown, and occur on the margins of the cups. Many species do not produce cups, and their podetia may be slender wands. *Cladonia* is an interesting but difficult group that often requires chemical analysis for identification.

29. **Dimelaena (Rinodina) oreina** [of mountains]. One of the most abundant crustose lichens of rock outcrops and cliffs. Its thallus covers and colors rocks throughout mountainous areas, and the greenish cast can be seen for miles. The apothecia are small, black, and imbedded in the thallus surface. It is one of the most common components of commercial "moss-rock."

30. **Lecanora argopholis** [silver-spot]. A lumpy, yellow-green or sometimes white crust, not lobed, with apothecia having a rim the same color as the thallus. The disk is a rich brown color. *Lecanora* is a very big genus of lichens that grow on rocks or bark, on the rotting stems of mosses, on dry wood, or even on bone. With experience with a few species, one can get a "feel" for the genus, although microscopic study is needed to be sure.

31. **Lecanora garovaglii** [for Santo Garovaglio, Italian botanist, 1805–1882]. This *Lecanora* is easily recognized by the fact that it has strongly developed marginal lobes which are very convex, as if inflated. The edges of the lobes are usually blue-black, and the apothecia are brown, with a rim the same color as the thallus (we say the rim is thalline or lecanorine). It is abundant on sandstones.

32. **Lecanora muralis** [of walls]. This *Lecanora* forms very neat little round thalli, much smaller than those of *L. garovaglii* and with the marginal lobes quite flat instead of rounded. The apothecia are tan to brownish. Common on sandstone.

33. **Lecanora novomexicana** [of New Mexico]. This *Lecanora* has a thick, shiny green thallus with short, convex marginal lobes. The apothecia are pale tan in specimens from low altitudes, but on the tundra rocks at high altitudes the apothecial disk is pruinose (with a white dusting) and the disk, when naked, is more blue-black.

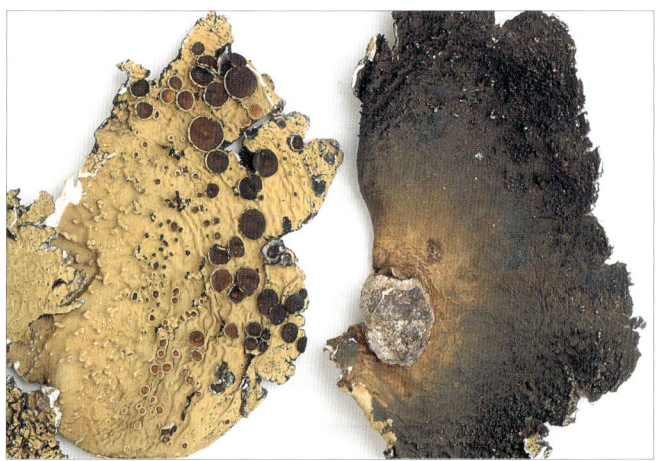

34. **Omphalora arizonica** [of Arizona]. This might aptly be called a potato chip lichen. On the rocky summits of the Colorado Rocky Mountains and desert ranges of New Mexico and Arizona this magnificent umbilicate lichen covers rocks with thalli as large as a foot in diameter! The thallus is bright green above, from pale brown to jet black below, and the surface is sharply ridged.

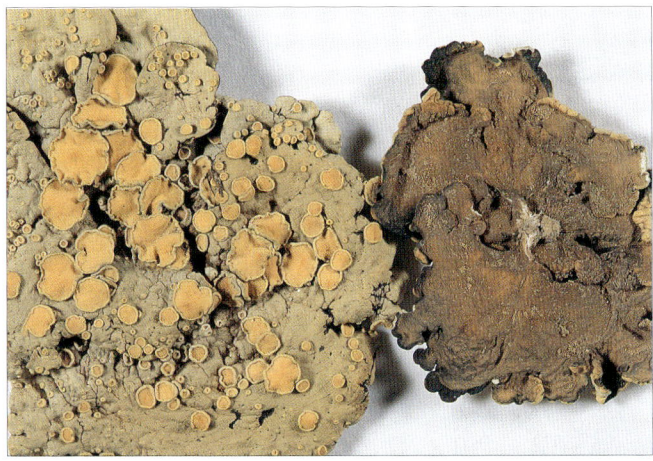

**35. Rhizoplaca (Lecanora) chrysoleuca** [yellow-white]. A curious lichen that is really crustose or squamulose, but has a lower cortex and a central umbilicus! The apothecia are orange or yellow-orange, and the thallus is rigid and can easily be removed from the rock with a penknife. In very sunny places the apothecia are brightly colored, paler in the shade. Very common on granitic rocks. Formerly known as *L. rubina*.

**36. Rhizoplaca (Lecanora) melanophthalma** [black-eyed]. Very similar to *R. chrysoleuca* and often growing alongside it on granite boulders. Here the apothecia may be pale greenish-gray, darkening to almost black. *Rhizoplaca* species differ from other Lecanora-like lichens by being umbilicate rather than closely attached to the substrate by the entire underside. The lower cortex is often black, especially toward the margins.

37. **Flavoparmelia (Parmelia) caperata** [wrinkled]. A very common, large, pale yellow, foliose lichen with broad, rounded lobes and a wrinkled surface covered with masses of soredia. This most commonly grows on vertical cliffs, but occasionally occurs on bark. Apothecia are hardly ever present.

38. **Flavopunctelia (Parmelia) flaventior** [more yellowish]. A fairly common large, pale yellow or more greenish lichen with broad, rounded lobes and with localized small patches of soredia. The surface of the thallus has scattered breaks in the cortex revealing the white medulla (pseudo-cyphellae), hence the epithet *punctelia* (spots). Usually this grows on bark, especially of piñon pines, but occasionally on rock. A close relative, *F. soredica*, has soredia only along the margins and lacks pseudocyphellae.

39. **Peltigera aphthosa** [plentiful]. *Peltigera* is a foliose lichen with peculiar apothecia that occur on the ends of the lobes, looking like brown tongues. Most *Peltigera* have blue-green alga hosts, but this one has green algae, hence the thallus is bright green when wet. But the thallus also has small black lumps on the upper side that are colonies of blue-green algae called *cephalodia*. So this is a symbiosis involving at least three different symbionts!

40. **Physconia (Physcia) muscigena** [growing on mosses]. A foliose lichen intermediate in size between the small *Physcia* types and the large *Parmelia* types, usually occurring in mats of mosses on rocks. The thallus is gray or brown, with a gray pruina, bright green when wet. There are no soredia. Apothecia are relatively uncommon, but when present usually are up to 5 mm broad and have a lobed thalline rim. The underside is black.

41. **Psora icterica** [jaundiced]. A soil lichen with a distinctive green thallus and black, convex apothecia. This grows in areas of heavy clay soil on the plains of Colorado, Texas, and New Mexico. It surprisingly reappears in southern South America!

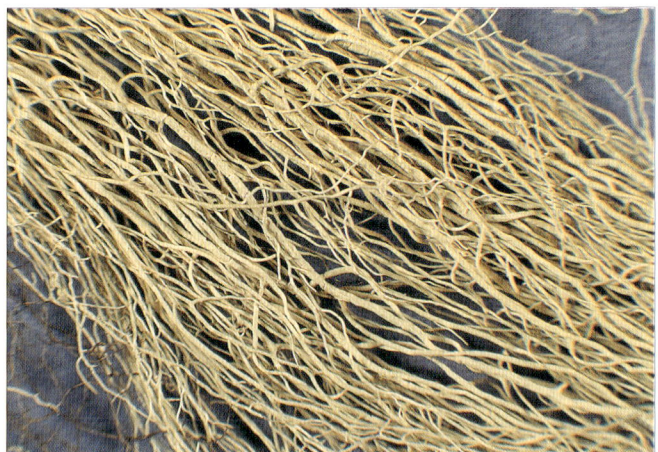

42. **Usnea cavernosa** [with cavities]. A long hanging species with a thallus often over a foot long, on conifers in the most moist sites in the mountains, often where fog hangs along the sides of the slopes. The thallus is smooth, without soredia, and the thicker branches have characteristic depressions or dents in the surface.

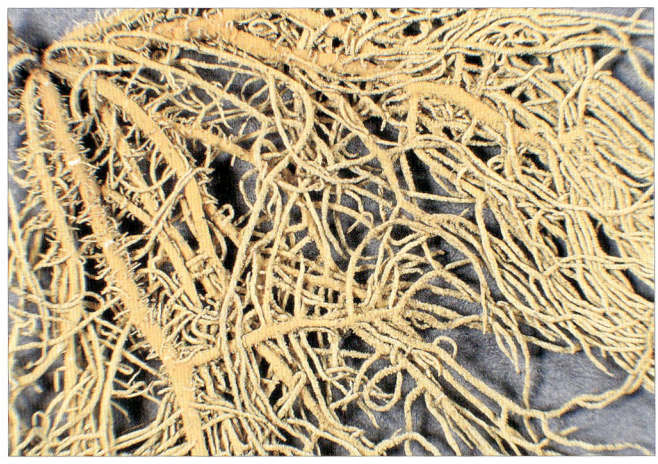

43. **Usnea hirta** [shaggy]. This is a small species of *Usnea*, usually found on the trunks of Douglas-fir trees. The main branches are short and stiff, and the entire thallus is studded with very short spinelike branchlets. There are no soredia. If the lichen is moist, gently tugging on each end of a branch will break the cortex, and reveal a stout white cord (the innermost part of the medulla). This feature separates *Usnea* from similar genera, such as *Alectoria* and *Ramalina*.

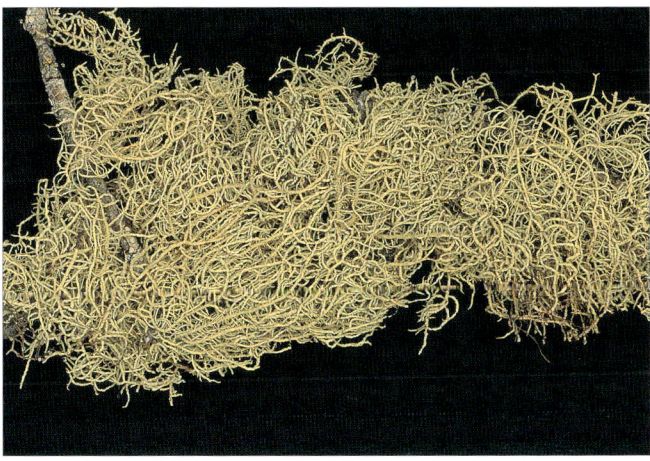

44. **Usnea lapponica** [of Lappland]. Probably the most common species, with a short, widely branched thallus of which the branch tips erupt with patches of soredia. Species of *Usnea* in the arid West require standing water in the tiny crevices of bark, and their presence usually indicates a humid microhabitat or the seasonal presence of a snow bank beneath the tree. The species appear to be short lived, and after possibly five years of growth, may suddenly disappear.

45. **Vulpicida (Cetraria) pinastri** [of pines]. A gray (green when wet) foliose lichen with masses of yellow soredia on the upturned margins, common at the bases of conifers in the foothills. It never fruits. The name *Vulpicida* has been given to lichens, some of which in olden times were mixed with meat used as bait to poison foxes and wolves.

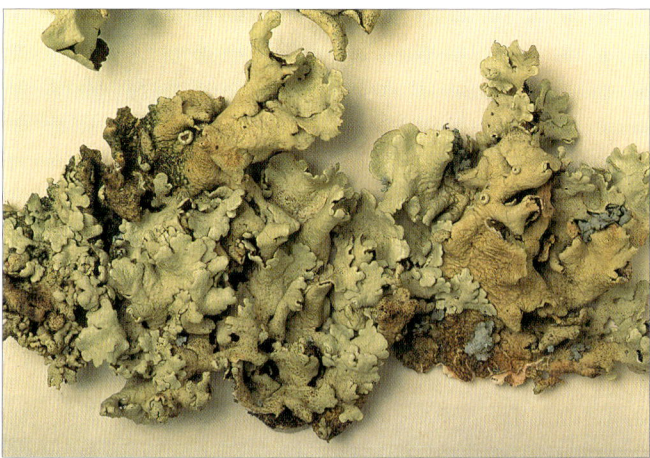

46. **Xanthoparmelia (Parmelia) coloradoënsis** [from Colorado]. The "green parmelias" are abundant and diverse in all mountainous and desert parts of the world. They are so abundant that they are actually responsible for the greenish cast common to the mountains themselves. The species are only distinguished by chemical analyses. *X. coloradoënsis* has no isidia or soredia, and forms a thallus loosely attached to rocks; the underside is pale.

47. **Xanthoparmelia (Parmelia) mexicana** [from Mexico]. This "green parmelia" is densely isidiate, and the underside is pale brown to brown. The apothecia of *Xanthoparmelia* are large, brown, with a thalline margin. What we call *X. mexicana* is probably a complex of several species or at least chemical strains.

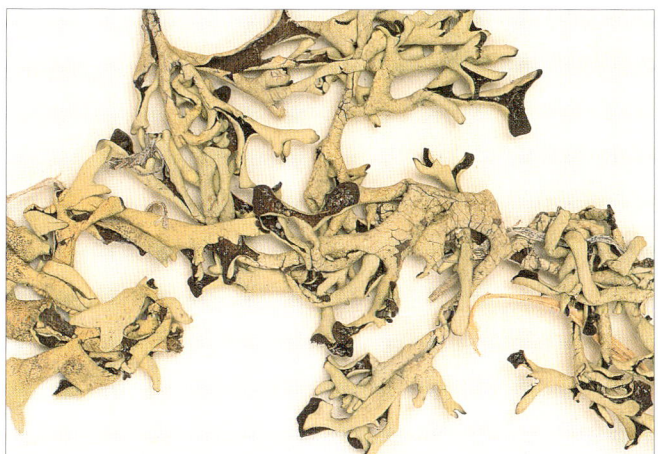

48. **Xanthoparmelia (Parmelia) vagans** [wandering]. An unattached foliose lichen with long lobes that are convolute (rolled from edge to edge). Apothecia are unknown. This plant rolls around free on the ground of prairies and open grasslands, often collecting in windrows. It occurs by the thousands and can be scooped up by handfuls. It is supposedly toxic to sheep. Possibly identical to *X. chlorochroa*.

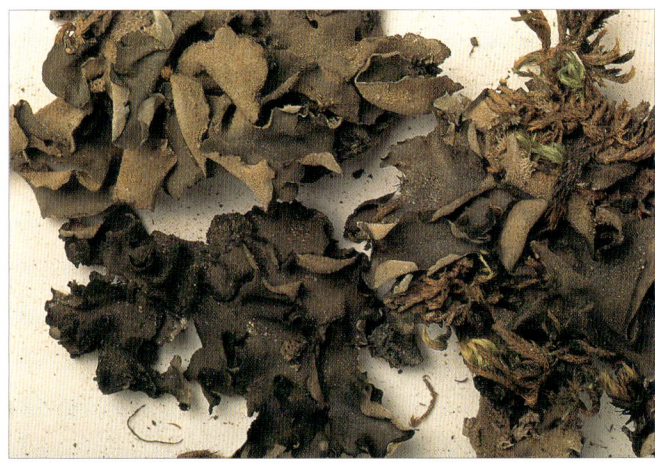

49. **Leptogium saturninum** [lead-colored; while the God Saturn is usually linked with gold, he is sometimes also associated with lead!]. A black, thin and softly foliose lichen characteristically found on seeping walls of vertical cliffs. The algal host is a blue-green alga, *Nostoc* (now more correctly called a cyanophilic bacterium). The underside is covered by white rhizinae. Foliose black lichens on wet rocks usually belong to either *Leptogium* or *Collema*.

50. **Pseudephebe (Parmelia) minuscula** [tiny]. An intensely black lichen composed of narrow, elongate, overlapping branches that are irregularly swollen and becoming almost threadlike at the tips, common on rocks of the alpine tundra. A second species, *P. pubescens*, is similar but the branches are much more slender and lack the irregular swellings.

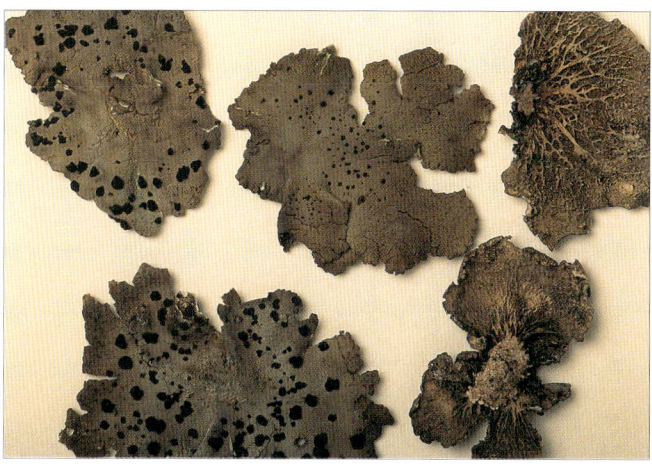

51. **Umbilicaria torrefacta** [of cataracts?]. Several small species of *Umbilicaria* are frequent on shaded granite boulders in the foothills canyons. They are dark brown or black, with circular or angular convex apothecia crisscrossed with deep furrows. This species has horizontal plates (called *trabeculae*) on the underside radiating from the central umbilicus. No other species has these.

52. **Aspicilia caesiocinerea** [blue-gray]. One of the most ubiquitous of lichens on granite rocks, covering large areas and giving the rock a gray color. The black apothecia are sunk in the thallus, giving the impression of molar teeth with black cavities. Very variable in thickness and lumpiness. Several species occur, requiring chemical and microscopic analysis for identification.

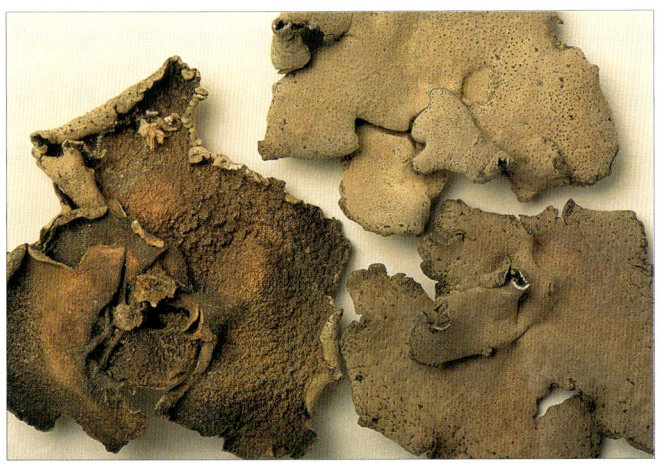

53. **Dermatocarpon miniatum** [bright red]. This umbilicate foliose lichen, when dry, has the tough texture of a fingernail. It grows on rocks that may be sporadically irrigated. The thallus is gray, with small black dots that actually are perithecia although one might think they are pycnidia. The underside is reddish brown, often somewhat wrinkled, and never with rhizines.

54. **Lasallia (Umbilicaria) pustulata** subsp. **papulosa** [with pustules; with small papillae]. A gray, umbilicate lichen characterized by having blisters on the upper cortex which are matched by corresponding indentations on the lower surface. It is abundant on shaded north faces of massive outcrops of granite in the outer foothills. The thallus is attached by a single umbilicus, and the underside has very fine papillae.

55. **Lecidea tessellata** [mosaic, tiled]. When well developed, this lichen is unmistakable, its black, flat apothecia on a level with the gray areoles, suggesting nothing less than a mosaic pavement, hence its name. Very common on sandstone and granitic rocks. In areas where the lichen is scoured by wind-blown sand or ice, the softer areoles are often eroded away, leaving a "ghost thallus" and the more dense and resistant apothecia standing alone.

56. **Parmelia saxatilis** [of rocks]. A gray foliose lichen with isidia covering the upper surface; the underside is black, and the rhizines are simple or only forked. It occurs in the same places as *P. sulcata*, but is not quite as common. Neither species ever has apothecia.

57. **Parmelia sulcata** [furrowed]. Probably the most common *Parmelia*, occurring mostly around the bases of conifer trees but also on rocks and over mosses. The thallus is gray, with thin cracks along the ridges; the underside is black, with black rhizinae that are stiffly horizontally-branched like a bottlebrush.

58. **Physcia stellaris** [star-shaped]. A rather small foliose lichen growing on the bark of various trees. The thallus is gray, with narrow convex lobes. The apothecia have black disks with a gray thalline margin. *Physcia* is a genus with many species, a few on bark but most of them on rocks. With a little experience, the genus can be recognized by its aspect (or *Gestalt*).

59. **Pseudevernia intensa** [intense]. A large, much-branched, tough, lichen with long lobes with convex, inrolled branches. The underside is black, and the upper surface usually has many small black dots (the pycnidia). Apothecia are very large with dark brown disks and gray thalline margins; they are elevated on short branches. The lichen is common on dead branches of conifers.

60. **Punctelia (Parmelia) hypoleucites** [pale underneath]. This large gray foliose lichen is almost identical to *Flavopunctelia* except that its thallus lacks usnic acid, which would color the cortex green. The pseudocyphellae are usually conspicuous, and the underside is pale tan. Several species of *Punctelia* occur, some with soredia, some with isidia, some with black undersides.

61. **Rhizocarpon geminatum** [paired, referring to the spores which occur two in an ascus]. A very common "salt-and-pepper" lichen found on granitic rocks throughout the mountains. This species has a very close relative, *R. disporum*. Actually, the names mean the same, two- or twin-spored. The spores are black, divided into many small cells, and in *R. disporum* they occur one to an ascus, belying the name; in *R. geminatum* they occur two to an ascus.

62. **Umbilicaria virginis** [of a maiden?]. A medium-sized white or pale tan species characteristically found on tundra boulders but occasionally on canyon walls, recognized by its pale underside covered with pale rhizinae. The apothecia are not furrowed (*gyrodiscs*) as in most *Umbilicaria*, but simply have a central invagination thus resembling a belly button (called an *omphalodisc*).

63. **Diploschistes scruposus** [rough stony]. Forms a uniform grayish or whitish crust on rocks and, in the alpine tundra, on soil. Immediately recognizable by the sunken, black, craterlike apothecia that appear to have "spokes" along the edges. Another species, *D. bryophilus* grows over the basal squamules of *Cladonia* lichens in the forests.

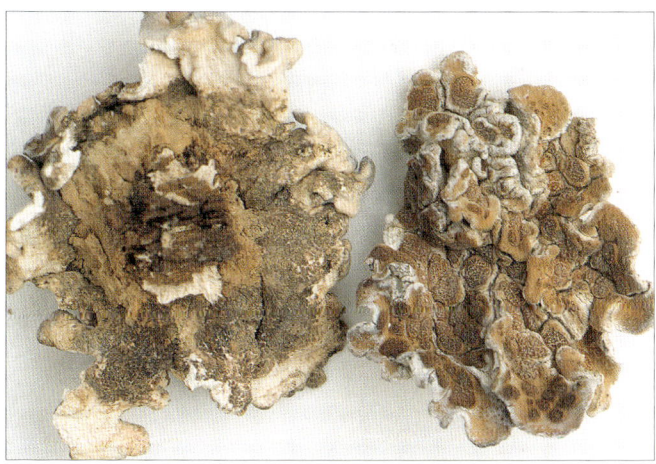

64. **Glypholecia scabra** [rough, scabby]. A curious lichen of cliffs in the desert canyons and plateaus. The thallus is thick and chalky white, but what we would call the crust is attached to the rock by a central umbilicus. It can be pried loose by carefully inserting a penknife under the thallus. The apothecia are brown and appear to be in groups, and their surfaces are criss-crossed by pale bands of sterile tissue.

65. **Icmadophila ericetorum** [of heaths]. Thallus a pale gray or white sheet overgrowing soft (punky) rotting wood in moist forests. The thallus is undifferentiated, not composed of discrete areoles. The apothecia are pink! There is nothing else like this. When well developed, the thallus can be a foot or more in diameter.

66. **Leprocaulon gracilescens** [slender]. An extremely delicate and brittle, slender white fruticose lichen with little knotty branches, growing in more or less tightly packed groups in crevices of granite rocks. Apothecia are unknown. A similar, smaller species with a distinctly green thallus is *L. microscopicum*, in similar sites. A third species, *L. subalbicans*, is white, and grows in tufts of mosses in the alpine tundra.

67. **Lepraria cacumina** [of the extreme point = Ultima Thule?]. A very abundant, coarsely lumpy white lichen dominating the soil in late-melting alpine snow-bed areas. This is a lichen that has never been known to fruit (what we call lichenes imperfecti). Several other species occur in forests, often at the bases of trees and sometimes on rocks. Some are called *Leproloma*; they are difficult to distinguish without chemical analysis.

68. **Normandina pulchella** [pretty, little]. This tiny and very unusual lichen resembles a little bluish-gray saucer, and can be found, usually over mosses, on the bark of the bases of conifers. The thallus is a single concave disk. This is one of our very few lichens that are not Ascomycetes; they belong to the Basidiomycetes, the group of fungi responsible for making mushrooms and toadstools.

69. **Psora cerebriformis** [brainlike]. On the desert soils of the Southwest this lichen forms conspicuous convex heaps of white convex thallus lobes interspersed with black convex apothecia. Desert soil lichens absorb moisture immediately if rained upon, and soften the force of rain. Also, during the winter, needle ice pushes up the surface layer of soil, with its lichens, mosses, and associated algae and fungi, forming a cryptogamic soil crust.

70. **Stereocaulon tomentosum** [woolly]. A white, fruticose lichen with elongate, branched stems (*pseudopodetia*) bearing clusters of scalelike structures (*phylloclades*). The main stem is woolly with matted hyphae. Close examination will reveal extremely small spheres hidden among the woolly hyphae; these are cephalodia containing blue-green algae. Brown or black, convex apothecia are often abundant, at or near the tops of the branches. Common on forest floors.

71. **Thamnolia vermicularis** [worm-like]. A very common, unattached lichen gathering in blowout patches of tundra or caught in between the stems of grasses and sedges. The thalli resemble hollow, white icicles, and may be simple or slightly branched. Apothecia are unknown. Two species are often recognized that are indistinguishable except that one fluoresces in ultraviolet light.

72. **Umbilicaria americana**. A huge umbilicate lichen up to 8 inches in diameter with a white or grayish thallus and jet black underside covered with black rhizinae, abundant on vertical cliffs on north-facing exposures in the foothills. Minute black dots on the thallus are pycnidia. Apothecia are very rare, but when present they are convex, black, and deeply furrowed. Long mistaken for *U. vellea*, this lichen was first recognized as distinct in 1994!

# ADDITIONAL SOURCES

Hale, Mason E. 1962. *Lichen Handbook.* 178 pp. Smithsonian Institution, Washington.

——— . 1967. *The Biology of lichens.* 176 pp. Wm. Clowes & Sons, Ltd., London.

——— . 1979. *How to Know the Lichens.* Wm. C. Brown, Dubuque, IA.

Hansen, Eric Steen. 1995. *Greenland Lichens.* 124 pp. Rhodos. Most of the lichens illustrated (in color) occur in the Rocky Mountains.

McCune, Bruce, & Linda Geiser, 1997. *Macrolichens of the Pacific Northwest.* 386 pp. Oregon State University Press, Corvallis.

McCune, Bruce, & Trevor Goward. 1995. *Macrolichens of the Northern Rocky Mountains.* 208 pp. Mad River Press.

Moberg, Roland. 1982. *Lavar* [Lichens]. 240 pp. Interpublishing. In Swedish, but the color illustrations are grand.

Richardson, David. 1975. *The Vanishing Lichens.* 231 pp. David and Charles, Vancouver.

Ursing, Björn. 1982. *Svenska Växter. Kryptogamer.* 530 pp. Norstedts, Stockholm. In Swedish. Fantastic watercolors of lichens, ferns, algae, and fungi.

Vitt, D. H., J. E. Marsh, & R. B. Bovey. 1988. *A photographic field guide to the mosses, lichens, and ferns of northwest North America.* 296 pp. Lone Pine Publishing, Edmonton.

Wirth, Volkmar. 1995. *Die Flechten Baden-Württembergs* [The lichens of Baden-Württemberg]. 2 volumes, 1006 pp. In German. The most elegant treatment of Central European lichens, with stunning color plates. Don't let the language put you off!

# GLOSSARY

**Alga** A green or blue-green one-celled plant (or cyanobacterium). In lichens these occupy a layer just beneath the cortex. (Also called phycobionts).

**Apothecium** (plural, **apothecia**) The most conspicuous spore-bearing structure on the surface of lichens; a disk-shaped structure that produces ascospores.

**Appressed** Closely anchored to the substrate.

**Areolate** Characterized by having the thallus broken into small, discrete patches or blocks.

**Ascocarp** A fungal structure that produces asci and ascospores (whether a perithecium or apothecium).

**Ascomycetes** A class of fungi in which spores are produced in asci; most lichen fungi are Ascomycetes.

**Ascospore** A spore produced by the apothecium of a lichen; these spores are produced initially by meiosis, in eights or multiples of eight, sometimes reduced to as few as two or one.

**Ascus** (pl. **asci**) A balloonlike cell which generates ascospores internally.

**Cephalodium** (pl. **cephalodia**) A separate colony of algae different from the normal symbiont, growing on the surface of a lichen.

**Cortex** A layer of tightly packed fungal tissue forming the upper or lower side of the thallus; sometimes the lower cortex is absent, exposing the medulla.

**Crustose** Closely attached to the substrate, and lacking free or ascending lobes.

**Cyphella** (pl. **cyphellae**) A break on the lower cortex, in which the medulla is not exposed but covered by a smooth membrane.

**Foliose** Having thin, leaflike lobes, loosely attached to the substrate.

**Fruticose** Having an erect, simple or branched, cylindrical thallus attached basally to the substrate.

**Fungus** (pl. **fungi**) An enormous class of lower plants characterized by lack of chlorophyll; includes mushrooms, toadstools, rusts, and smuts.

**Genus** (pl. **genera**) A group of closely related species having a common line of descent.

**Habitat** The milieu in which a lichen occurs. Common habitats are rock (the lichen is saxicolous), bark (the lichen is corticolous), wood (the lichen is lignicolous), mosses (the lichen is muscicolous), and soil (the lichen is terricolous).

**Hypha** (pl. **hyphae**) White, threadlike chains of fungal cells forming a cottony mass.

**Isidium** (pl. **isidia**) A special growth on a lichen thallus, usually spherical or cylindrical, which contains a bit of medulla surrounded by cortex and which is brittle when dry and may break off and propagate another lichen thallus.

**Lobe** A marginal portion of a thallus, normally rounded, scalelike or leaflike.

**Margin, thalline** The rim of an apothecium, if it is the same color and texture as the thallus (also sometimes called a lecanorine margin).

**Margin, proper** The rim of an apothecium, if it is the same color and texture as the disk (also called a proper exciple). An apothecium may have both a thalline and a proper margin.

**Medulla** (pl. **medullae**) An undifferentiated, cottony mass of fungal hyphae forming the internal portion of a lichen thallus.

**Papilla** (pl. **papillae**) Minute discrete bumps on the thallus surface.

**Perithecium** (pl. **perithecia**) A spherical ascocarp completely or partially sunken in the thallus.

**Podetium** (pl. **podetia**) The erect branches of *Cladonia* thalli.

**Pruina** (adjective, **pruinose**) A gray or white, waxy, or powdery covering of parts or all of a thallus; often this consists of crystals of calcium oxalate precipitated by the lichen following absorption of calcium from the substrate.

**Pseudocyphella** (pl. **pseudocyphellae**) A break in the thallus which exposes the medulla; this may be extremely small, a mere point, or may cover a fairly large area.

**Pycnidium** (pl. **pycnidia**) A structure producing special spores, called conidia or pycnoconidia (functions unclear), produced by vegetative budding; pycnidia appear as small black dots on the surfaces of many lichens.

**Rhizine** (pl. **rhizinae**) Hairlike or bristlelike structures growing on the margins or lower surfaces of lichens; these may or may not function as anchors to the substrate.

**Soralia** Aggregates of soredia.

**Soredium** (pl. **soredia**) Minute powdery or granular clusters of medullary hyphae and algal cells, produced on lichen surfaces, easily detached, and serving to propagate new thalli.

**Species** A particular organism or population with distinct characteristics, and having a common line of descent. Species are real, although botanists have difficulty providing a universal definition, some claiming that a species is what a competent taxonomist considers to be one.

**Spore** A one- to many-celled unit serving a reproductive function in fungi.

**Squamule** (adjective, **squamulose**) A scalelike lobe or thallus occurring, for example in some *Psora* species.

**Substrate** The material to which a lichen is attached.

**Symbiont** One of the partners of the fungal/algal relationship of lichens.

**Thallus** (pl. **thalli**) the body of a lichen.

**Umbilicus** (adjective, **umbilicate**) A restricted attachment point, often at the center of a thallus, anchoring it to the substrate.

# INDEX

*synonyms appear in italics*

| SPECIES | PLATE |
|---|---|
| *Acarospora chlorophana* | 1 |
| *Alectoria chalybeiformis* | 21 |
| Aspicilia caesiocinerea | 52 |
| Bryoria chalybeiformis | 21 |
| Caloplaca trachyphylla | 11 |
| Candelariella rosulans | 2 |
| Cetraria aculeata | 22 |
| *Cetraria cucullata* | 4 |
| Cetraria islandica | 16 |
| *Cetraria nivalis* | 5 |
| *Cetraria pinastri* | 45 |
| *Cetraria tilesii* | 6 |
| Cladonia cariosa | 24 |
| Cladonia cervicornis var. verticillata | 25 |
| Cladonia deformis | 26 |
| Cladonia ecmocyna | 27 |
| Cladonia pyxidata | 28 |
| *Cladonia verticillata* | 25 |
| Coelocaulon aculeatum | 22 |
| Dactylina madreporiformis | 23 |
| Dermatocarpon miniatum | 53 |
| Dimelaena oreina | 29 |
| Diploschistes scruposus | 63 |
| Flavocetraria cucullata | 4 |
| Flavocetraria nivalis | 5 |
| Flavoparmelia caperata | 37 |
| Flavopunctelia flaventior | 38 |
| Glypholecia scabra | 64 |
| Icmadophila ericetorum | 65 |
| Lasallia pustulata subsp. papulosa | 54 |
| Lecanora argopholis | 30 |

# Index

| SPECIES | PLATE |
|---|---|
| *Lecanora chrysoleuca* | 35 |
| Lecanora garovaglii | 31 |
| Lecanora melanophthalma | 36 |
| Lecanora muralis | 32 |
| Lecanora novomexicana | 33 |
| *Lecanora rubina* | 35 |
| Lecidea atrobrunnea | 15 |
| Lecidea tessellata | 55 |
| Lepraria cacumina | 67 |
| Leprocaulon gracilescens | 66 |
| Leptogium saturninum | 49 |
| Letharia vulpina | 7 |
| Melanelia subolivacea | 17 |
| Nephroma parile | 18 |
| Normandina pulchella | 68 |
| Omphalora arizonica | 34 |
| *Parmelia caperata* | 37 |
| *Parmelia chlorochroa* | 48 |
| *Parmelia coloradoënsis* | 46 |
| *Parmelia flaventior* | 38 |
| *Parmelia hypoleucites* | 60 |
| *Parmelia mexicana* | 47 |
| *Parmelia minuscula* | 50 |
| Parmelia saxatilis | 56 |
| *Parmelia subolivacea* | 17 |
| Parmelia sulcata | 57 |
| *Parmelia vagans* | 48 |
| Peltigera aphthosa | 39 |
| Peltigera rufescens | 19 |
| Peltigera venosa | 20 |
| Physcia muscigena | 40 |
| Physcia stellaris | 58 |
| *Physconia muscigena* | 40 |
| Pleopsidium chlorophanum | 1 |
| Pseudephebe minuscula | 50 |
| Pseudevernia intensa | 59 |
| Psora cerebriformis | 69 |
| Psora decipiens | 12 |
| Psora icterica | 41 |

# Index

| SPECIES | PLATE |
|---|---|
| Psoroma hypnorum | 14 |
| Punctelia hypoleucites | 60 |
| Rhizocarpon geminatum | 61 |
| Rhizocarpon geographicum | 3 |
| Rhizoplaca chrysoleuca | 35 |
| Rhizoplaca melanophthalma | 36 |
| *Rinodina oreina* | 29 |
| Solorina crocea | 13 |
| Stereocaulon tomentosum | 70 |
| Thamnolia vermicularis | 71 |
| Umbilicaria americana | 72 |
| Umbilicaria pustulata subsp. papulosa | 54 |
| Umbilicaria torrefacta | 51 |
| Umbilicaria virginis | 62 |
| Usnea cavernosa | 42 |
| Usnea hirta | 43 |
| Usnea lapponica | 44 |
| Vulpicida pinastri | 45 |
| Vulpicida tilesii | 6 |
| Xanthoparmelia coloradoënsis | 46 |
| Xanthoparmelia mexicana | 47 |
| Xanthoparmelia vagans | 48 |
| Xanthoria elegans | 8 |
| Xanthoria fallax | 9 |
| Xanthoria montana | 10 |

**DATE DUE**

DUE DATE SUBJECT TO CHANGE
IF A RECALL IS REQUESTED